THE
WORLD'S
GREATEST
PAPER
AIRPLANE
AND
TOY
BOOK

For Matthew

THE
WORLD'S GREATEST
GREATEST
PAPER
AIRPLANE
AND
TOY
BOOK

KEITH R. LAUX

TAB Books
Division of McGraw-Hill

New York San Francisco Washington, D.C. Auckland Bogotá
Caracas Lisbon London Madrid Mexico City Milan
Montreal New Delhi San Juan Singapore
Sydney Tokyo Toronto

Special thanks to Jeff Long
for his technical assistance
in testing the plane designs

© 1987 by **TAB Books**.
TAB Books is a division of McGraw-Hill, Inc.

24 25 26 27 28 29 BKM BKM   0 9 8

ISBN-13: 978-0-8306-2846-9

ISBN-10: 0-8306-2846-0

Library of Congress Cataloging-in-Publication Data

Laux, Keith R.
  The world's greatest paper airplane and toy book / by Keith R. Laux.
      p.   cm.
    Includes index.
    ISBN 0-8306-2846-0 (pbk.)
    1. Paper airplanes. I. Title.
TL778.L38 1987
745.592—dc19   87-19418
                                                     CIP

Contents

Introduction vii

Folding Hints viii

Flying Hints x

Part 1: Competition Craft 1

The Master 2
The Standard 4
Folded-Nose Box Plane 6
Hog-Nosed Plane 8
Shibumi 10
Undercarriage Plane 12
The Looper 14
The Sprinter 16
Origami-Nose Box Plane 18
The Dragon Bird 20
Sail Wing 24
Saber-Toothed Bat 26
The Gull 28
The Stunt Looper 30
Lock-Back Jackknife 32
The Fly 34
The Gremlin 36
The Hunter 38
Ram Jet 40
The Boat Plane 42
The Wing 44
Super Dart 46
Bee Bomber 48

X-Wing Fighter   50
King Wing   52
Glider   54
Hawk   56
Super-Modified Box Plane   58
SST   60
Long-Distance Champ   62

Part 2: Novelty Planes   65

Tail Plane   66
Modified Tail Plane   68
Spin King   70
The Circular Airfoil   72
Flat Flyer   73
Major Arrow   74
Helicopter   77
Eagle   78
Dove   80
Dragonfly   82

Part 3: Toys, Toys, Toys   87

The Box   88
Frog   90
Pac-Man Mouth   92
Hats   95
The Fortune Teller   96
Fox Face   98
Bird   100
Star   102
Pinwheel   104

By trade, I am an engineer, but quite often the child within me sneaks out in search of amusement and playful diversion. At times, I turn to a ball and bat; at other times, to a house of cards. Even though such activities are enjoyable, far more often I pick up a sheet of paper and begin to carefully fold and crease. In so doing, I transform a simple sheet of a memo pad into a creation capable of something wonderful . . . *flight!*

With some slight effort and care, I have put down my thoughts, along with those of others, to give you a complete manual of the art of making paper airplanes. Every type of folding technique is illustrated in more than 40 different aircraft designs. Each plane is unique and uses unusual techniques that ensure a smooth flight. You may say "Obviously, there must be more than 40 different possible planes," and you are right! This book will give you the tools for literally hundreds of variations; thousands even, if you wish to pick up a pair of scissors to cut and modify the wings. Adding tail sections can give you even more possibilities.

This book is a guide to teach you. Please remember as you try the models in this book that your imagination can give each plane your own personal touch. Experiment with flaps and stabilizers as much as possible. The secret of successful flight lies in having the correct trim on your plane. Start by achieving a smooth, level flight, then adjust the lifting and directional surfaces for rolls and loops. Practice will reward you with a memorable flight.

So, for schoolchildren and businessmen alike, and for any of you who wish to learn more about an age-old art, read on, with paper in hand.

Because this is truly a book about folding paper airplanes, you will need no other tools to complete every model plane in this book than the following:

A sheet of paper
Your hands
Your mind

Not too difficult? O.K. You can use any kind of paper, but typing paper or an 8½- x -11-inch pad of paper will work best. To get technical, typing paper between 16 and 20 lb. is fine. All of the planes start with an 8½- x -11-inch sheet. Some require you to make a square sheet from it. Figure 1 shows you how to do this step. Simply fold and crease, then tear off the strip to get a square.

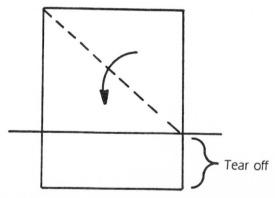

Tear off

It is important that you fold *and crease* each step of every model. The easiest way is to use your fingernail to push down each fold as you make it.

Reverse and inverted folds are used quite often in the designs. Figures 2 and 3 illustrate how to make them.

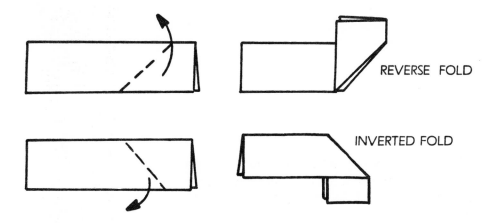

REVERSE FOLD

INVERTED FOLD

Each of the drawings are to scale and should look exactly like the plane you are folding. If the plane you have does not look right, try it first. It might work better than the intended design. If not, go back and try again.

Try heavier paper if you like. Most planes will fly well outdoors with heavy paper. Lighter paper is not recommended because it usually will not hold a crease very well.

Lastly, a note about glue, tape, and staples. You can use them on many of the models, but they are not necessary. If you would like to make sure the plane will not come apart, try them. Staples and tape are also good for adding extra weight if you want to experiment with weighting the planes to change their flight.

Making the planes is only about 63 percent of the total art of paper airplanes. The rest is in the flying. There are many things you can do to assist with the last 37 percent. They include making changes in these areas:

☆STABILIZERS☆

You can add stabilizers to any plane. They are used to help the plane fly straight or to control curves. You can add them to the wing edges, bending either up or down. You can also invert-fold the tails of many planes up through the top to get a center stabilizer. Angling them can influence the lift and drag of the plane.

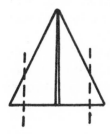  

Flying Hints

☆TABS AND BENDS☆

You can use tabs and bends to control the lift and drag of your aircraft. Many planes need the extra lift these control surfaces provide simply because the wing shape is never a very good airfoil on a paper plane. Some extra wing trim is good for a level flight on each plane in this book. You can use any method that you think is appropriate or aesthetic. If it works, it works.

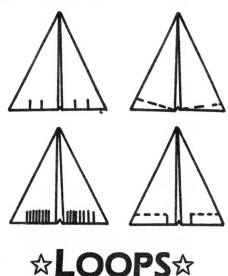

☆LOOPS☆

There are two basic ways to get your plane to do loops. a: Bend the tabs up for extra lift, and throw the plane hard and slightly down away from you. Some planes might just nose up and stall when you do this. Experiment. b: Hold the belly of the plane toward you and throw with medium force straight up. The plane should loop away from you and level out, flying back in your direction. Try to catch it for a hand-to-hand loop.

☆**CURVES**☆

You can use stabilizers or tabs to fly curves with your plane. Bending only one tab up or angling your stabilizers will work. Throw the plane with a light push.

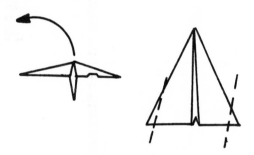

☆**ROLLS**☆

Bending one tab up and one tab down and throwing hard will usually give you rolls. In fact, most things that will hurt smooth, level flight will produce rolls if you throw hard enough.

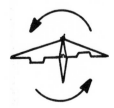

COMPETITION
CRAFT

The designs in this first chapter are all highly competitive craft. Many have won awards in various contests.

PART 1

☆THE MASTER ☆

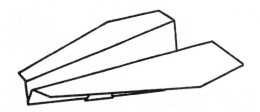

This first design is, logically, the one that is my personal favorite. I have won several contests with it and have been using this design since I was ten years old. It is easy to fold, uses sound aerodynamics, and is perhaps the most stable and graceful plane in this book. With a slight bend at the wing tips, this plane will perform amazing loops and rolls. It is also a long-range glider.

① Start with an 8½-x-11-inch sheet. Crease it in half lengthwise and open. Then fold the corners down as shown.

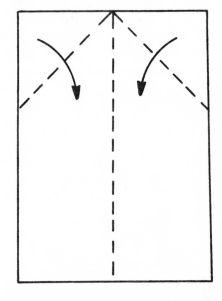

② Fold down the top section as shown.

③ Fold the top corners to the centerline so a slightly flattened diamond shape shows through at the center.

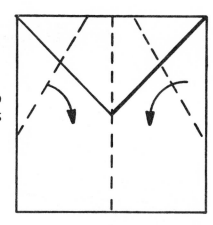

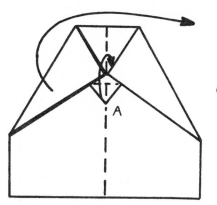

④ Fold point A up to lock in the two flaps. Now fold the plane in half, away from you.

⑤ Fold the wings out on each side at a slight angle, as shown. This angle increases stability greatly.

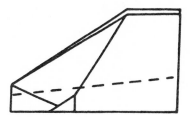

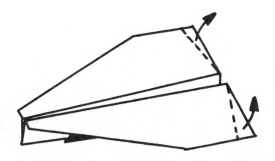

⑥ If lift is needed, bend the wing tips up as shown.

TIP: Because of the weight at the bottom, the plane is very stable. If you throw it straight up outdoors, it will flatten out and glide to earth. To obtain the best glide allow the wings to separate slightly.

☆ THE STANDARD ☆

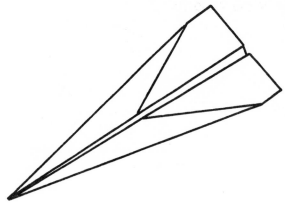

This model is here for the sake of nostalgia. Almost everybody has seen this plane, whether in the grade-school classroom or in the corner of someone's office. Perhaps the Japanese, with their ancient experience in paper folding, created the first paper plane. Western civilization surely recognizes this design as the grandfather of them all!

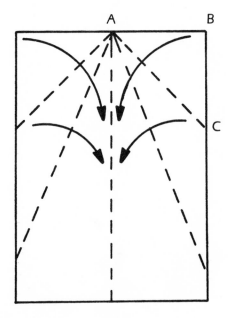

① Begin with an 8½- x -11-inch sheet. Fold it in half lengthwise and open again. Bring the top edges (A-B) to the center-line, as shown. Then, keeping it folded in, fold the edges (A-C) in to the center.

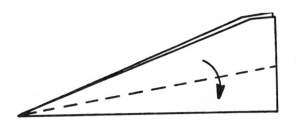

② Fold the plane in half toward you and fold out the wings along the line.

TIPS: To obtain an underhand loop, bend up the wing tips at a 30-degree angle and launch.

Improved Standard

For an Improved version of the Standard,

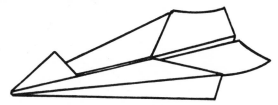

① Fold the sheet in half along both its length and width. Bring the corners down to the dots as closely as you can. As you press these folds down, you should have a flap in the middle.

② Fold the sides (A-B) to the center under the flap. Fold the plane in half toward you.

③ Now fold the wings out as for the Standard. The fin up front will add stability in flight.

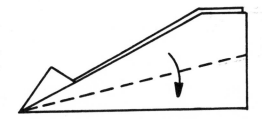

☆ FOLDED-NOSE BOX PLANE ☆

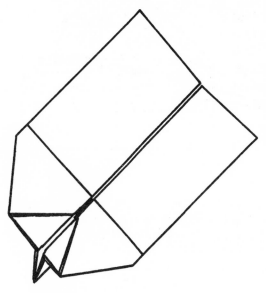

The box plane is another fond look back to the toys of childhood. For me, this plane was a mystery for quite awhile. Finally, I traded a bag of marbles for the secret of its design. Over the years, the box plane has proven to be a consistent performer; so good in fact, that a similar design won the time-aloft competition in the first Great International Paper Airplane Competition, held in New York in February 1967.

① Fold the corner of an 8½-x-11-inch sheet to the other side as shown.

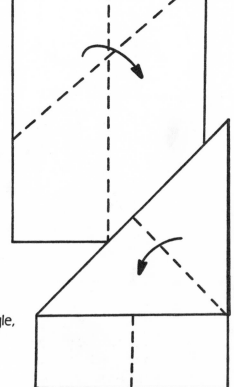

② Fold the top point over to get a triangle, then open both folds out again.

③ Push the sides in with an inverted fold and press the paper flat to get the triangle with accordian folds, as shown. (This fold will be used to begin many planes in this book.) Fold up the bottom corners to the top.

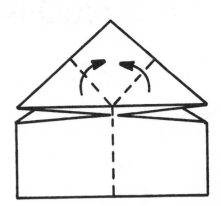

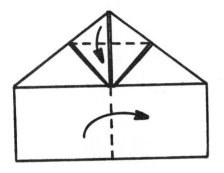

④ Fold over at the line, even with the two sides of the diamond. Now fold the plane in half toward you (with the folds inside the fold).

⑤ Fold the wings down as shown. Then fold down along the top line for stabilizers.

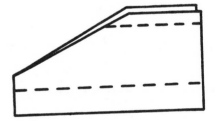

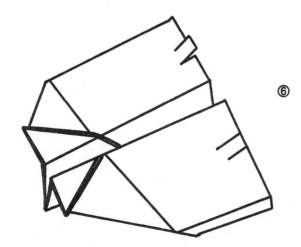

⑥ The completed plane. If lift is needed, tear small tabs up at the back of the wings and bend them up as much as necessary.

TIPS: With the tabs bent up, this plane will do nice loops. With one tab up and one down, the plane will do rolls and curves.

☆ HOG-NOSED PLANE ☆

The Hog-Nosed Plane is a simple variation of the standard model. The additional folds provide a better weight distribution for stunts. Its characteristics are much the same as those of the Box Plane, but the Hog-Nosed Plane can be thrown with much greater force.

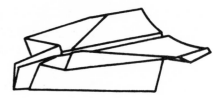

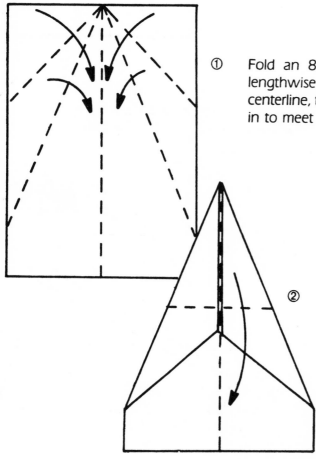

① Fold an 8½-x-11-inch sheet in half lengthwise. Fold the top sides in to the centerline, then bring the diagonal sides in to meet in the middle also.

② Fold the top point to the bottom of the sheet.

③ Fold the tip back up, making the crease just above the intersection of the other edges.

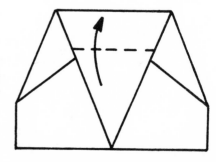

④ Now fold the last bit back down even with the previous crease. This fold will stick out a bit, like a hog's nose. Fold the plane in half toward you.

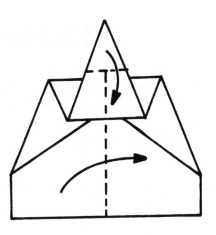

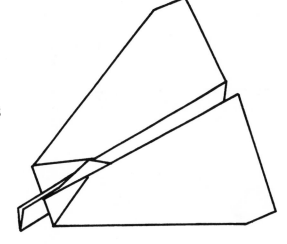

⑤ Fold out the wings and fold down the edges. The stabilizers can go either up or down, or can be left out if you wish.

⑥ The completed plane. You can add tabs to make this sky hog fly better!

TIPS: This plane flies well with a light send off. It will perform loops and rolls if you adjust the wing tabs.

☆ SHIBUMI ☆

This model shows the grace and form so highly valued in Japanese paper folding. *Shibumi* is a word that embodies the height of carefully restrained grace and form. I learned this design as a child and have caught the interest of many with it over the years. As with most origami designs, it is folded from a square sheet. Good luck.

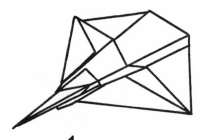

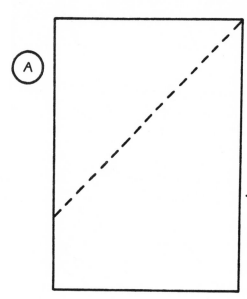

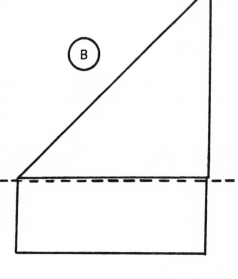

① Use an 8½-inch-square sheet of paper, made by folding a square from an 8½- x -11-inch sheet and tearing off the remaining strip. (See figures A and B). Fold the edges to the center, as shown.

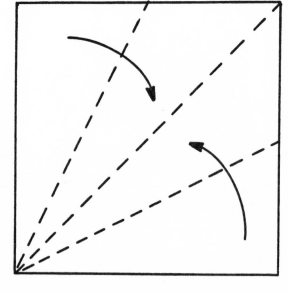

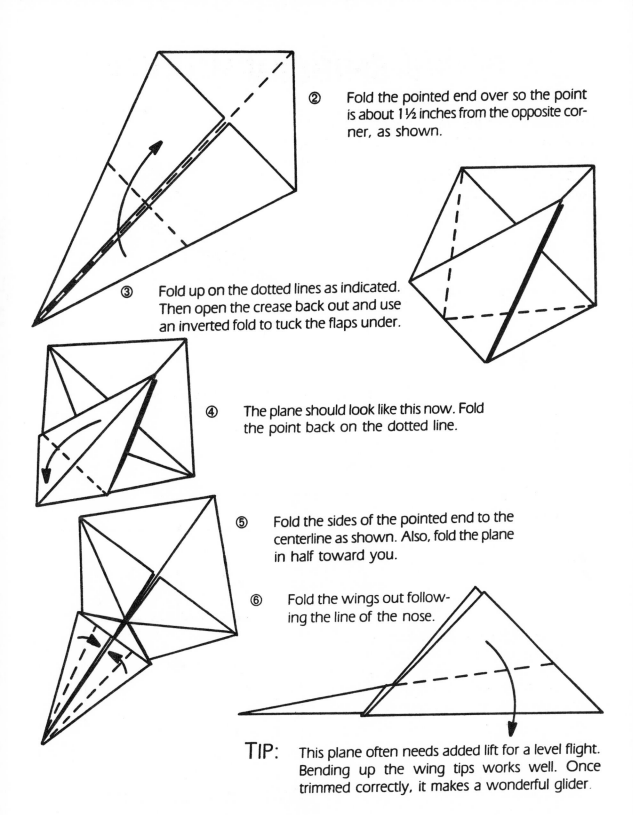

② Fold the pointed end over so the point is about 1½ inches from the opposite corner, as shown.

③ Fold up on the dotted lines as indicated. Then open the crease back out and use an inverted fold to tuck the flaps under.

④ The plane should look like this now. Fold the point back on the dotted line.

⑤ Fold the sides of the pointed end to the centerline as shown. Also, fold the plane in half toward you.

⑥ Fold the wings out following the line of the nose.

TIP: This plane often needs added lift for a level flight. Bending up the wing tips works well. Once trimmed correctly, it makes a wonderful glider.

11

☆ UNDERCARRIAGE PLANE ☆

Here is a different kind of glider. From the bottom, it almost looks as complex as a real jet. If you experiment, I'm sure you can make landing gear for it. There are some tricky folds here. After you see how well it glides, though, you will agree that the work was worth it. Take it slowly. You can use these folds on other designs, too.

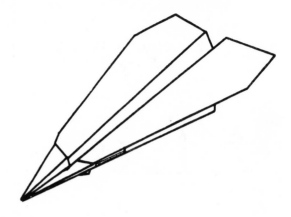

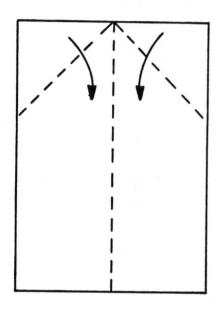

① Start with an 8½- x -11-inch sheet. Fold it in half lengthwise and open. Fold the corners to the center.

② Fold over at the line and crease.

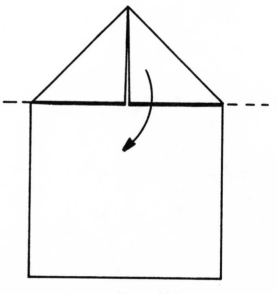

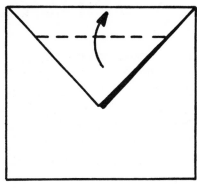

③ Now, leaving a little more than 1 inch, fold the flap back up.

④ This is the tricky part. From the tip of the top point, fold all the way to the bottom corners. Open; fold up along the other set of dotted lines as shown and unfold. Lift section A up, invert fold section A, and push section B underneath. Repeat on the other side.

⑤ Fold in corner A to the centerline; crease and unfold. Stick a finger in at the arrow and fold along the short line. As you do this, corner A should lift up. Bring A back to the center and invert fold at B back down on top. Repeat this for the other side.

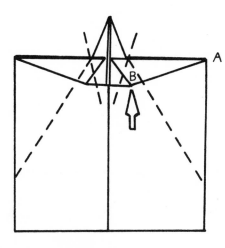

⑥ Fold the plane in half away from you. Then fold the wings out along the dotted line. You now have a long-range glider.

TIP: You can lift the wing tips slightly to help in a level flight. Because this plane is so long, it does not do loops, but you will be amazed by its long, straight glides.

☆ THE LOOPER ☆

The Looper is so named because of its ability to do hand-to-hand stunts, particularly loops. It has a canard-type wing up front that acts as a control surface to give it amazing stability. This is truly a plane that can be flown outdoors, even in heavy winds. It is no wonder that experimental aircraft with this front wing are, in some ways, revolutionizing the airplane industry.

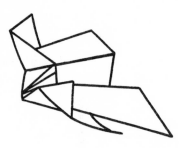

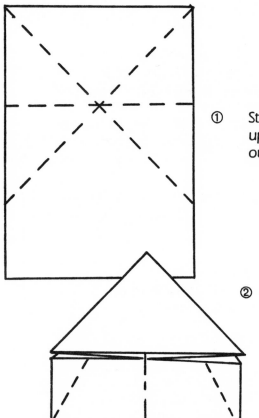

① Start with an 8½-×-11-inch sheet. Fold up on the dotted lines, crease, and open out again.

② Push the sides in to make an accordian fold, as for the Folded-Nose Box Plane. Now take the underneath surface and fold it in so the crease ends in the corners, as shown.

③ Divide the top triangle into thirds by eye. Fold at A toward you and at B away from you. Then fold the plane in half away from you.

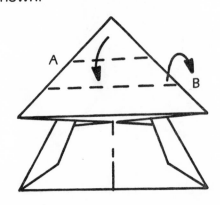

④ Fold the wings out at a fairly steep angle, as shown by the dotted line.

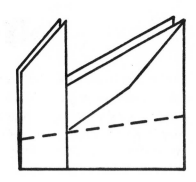

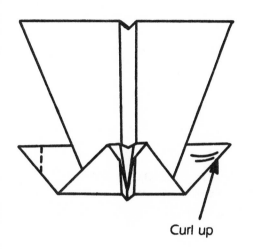

Curl up

⑤ A top view of the plane. Curl the front edges of the canard wing up to control the rate of looping. The more curl, the more loop. By only folding up the tips on the dotted line, you will have a stable glider.

TIPS: Experiment with tabs on the back wings. Also experiment with the front wing tips. You can coax rolls, loops, and curves from this winner.

☆THE SPRINTER☆

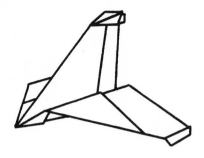

This plane has its own rudder and looks a lot like a racing jet. Sometimes the wings flap as it flies, sometimes not. Either way, this plane employs some very unique folds that you can adapt to planes of your own design. It flies quickly and can do some funny stunts. Experiment with this one!

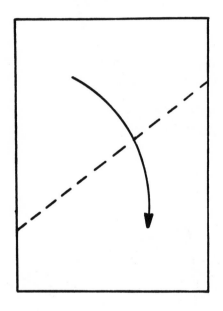

①   Fold an 8½- x -11-inch sheet so that the opposite corners meet, as shown.

②   Fold along dotted line A so the edge is even with the two side points. Then fold the plane in half, as shown (B).

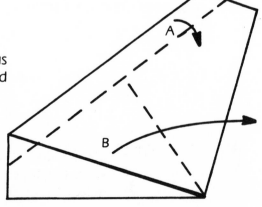

③ Fold the top sections down on both sides. Flip the folded paper over (180 degrees).

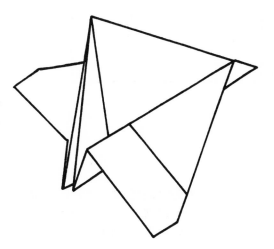

④ Open out the wings as shown (at a slight angle). Leave the center section as is.

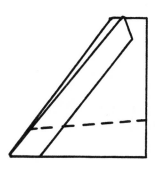

⑤ This is how the completed plane should look. You can staple or glue the bottom together if desired, for better flying.

TIPS: The Sprinter is a quick flier, and it might take a short while to get the trim adjusted correctly. Bend the wing tips up to level out its flight.

☆ ORIGAMI-NOSE BOX PLANE ☆

This is a more difficult box plane to construct, but its looks are worth it. The design incorporates a basic origami fold to weight the nose section. There are many ways to construct the nose of this plane. After reading the rest of this book, you should be able to adapt several other folds to this basic design. This plane is a good choice to use to get your message across.

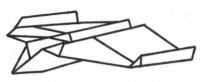

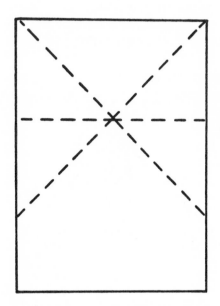

① Using an 8½-x-11-inch paper, start as for the Folded-Nose Box Plane. Fold up and crease on the dotted lines. Push in the sides and flatten (an accordian fold).

② Fold the bottom corners up to the top along the dotted lines.

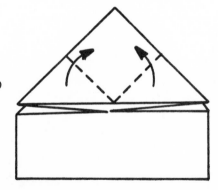

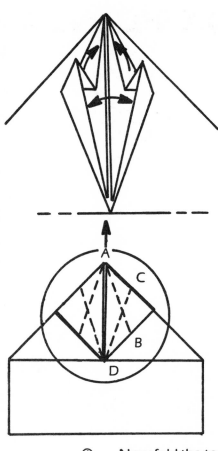

③   This is the tricky part. Fold the top two sides of the diamond into the center along line A-B and crease. Open and fold along line C-D, bringing the bottom sides to the center. Now follow the closeup. Push the top edges in and reverse-fold the paper so the bottom edges follow on top. There will be two little horns pointing up. Flatten them down.

④   Now fold the top away from you along the line. The two horns should stick straight out.

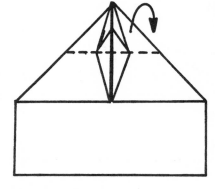

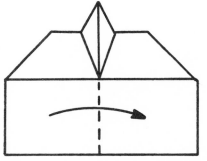

⑤   Fold the plane in half toward you.

⑥   Fold out the wings and fold down the stabilizers.

TIP: Add tabs to the wing tips and experiment to get the desired flight.

☆ THE DRAGON BIRD ☆

This plane looks like no other and uses some very unusual folds. I learned its design in California from a boy on a skateboard who was towing this creation by a kite string like a flying pet. This one truly will get attention flying across the classroom or the boardroom. Inquirers should be charged a pretty penny before you teach them this one. Experiment with the tail up and down; you can also reverse-fold the nose to make a beak for this bird!

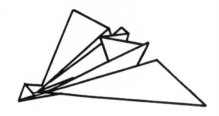

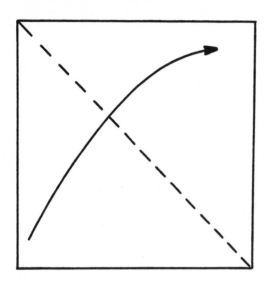

① Start with an 8½-inch-square sheet. Fold it in half as shown.

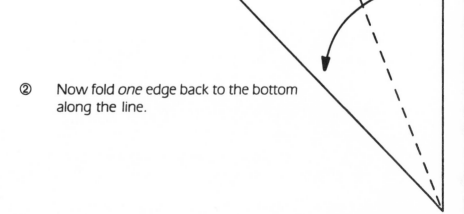

② Now fold *one* edge back to the bottom along the line.

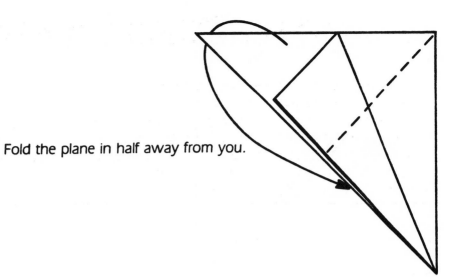

③ Fold the plane in half away from you.

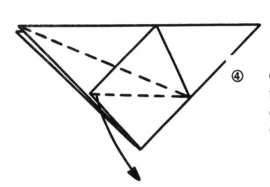

④ Crease along the line by folding the bottom left side to the top. Open back up, grasp the top layer, and pull it out and over to get figure 5. Flatten it down.

⑤ Open the top surface out as shown.

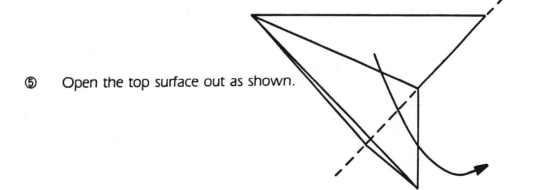

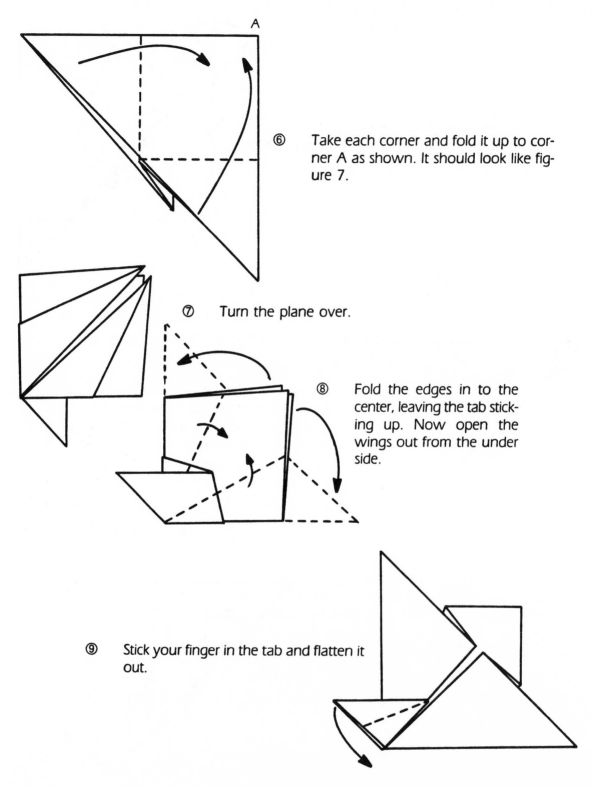

⑥ Take each corner and fold it up to corner A as shown. It should look like figure 7.

⑦ Turn the plane over.

⑧ Fold the edges in to the center, leaving the tab sticking up. Now open the wings out from the under side.

⑨ Stick your finger in the tab and flatten it out.

⑩ Turn the plane over and fold in half toward you.

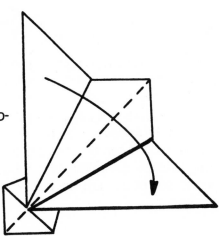

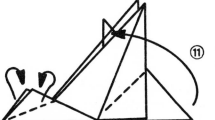

⑪ Tuck in the nose by folding in as shown. You can reverse fold the tail up if you like.

⑫ Fold the wings down along the lines as shown.

TIP: Bend the wing tips up for a level flight. The Dragon Bird glides well, but cannot do stunts. It is one of those all show and no go types you've heard about.

☆ SAIL WING ☆

The Sail Wing was an entry in the first Great International Paper Airplane Competition, but has been modified slightly to achieve better performance. It is very much like the box planes shown earlier, but without the folded wings. The nose is yet another variation, and can be used on the box plane as well. With the back lifting tab, the plane glides well with a soft push—just the right touch for a noisy auditorium.

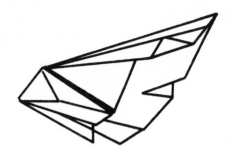

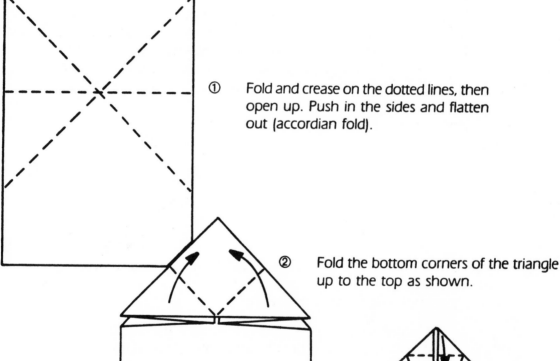

① Fold and crease on the dotted lines, then open up. Push in the sides and flatten out (accordian fold).

② Fold the bottom corners of the triangle up to the top as shown.

③ Fold the bottom sides of the diamond into the center. Now fold the top over as indicated.

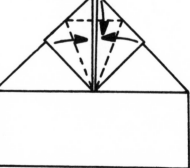

④ Open up the slots in the top section and slide the corners of the flaps into them to lock them together. Flatten the plane out.

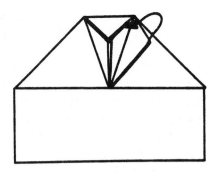

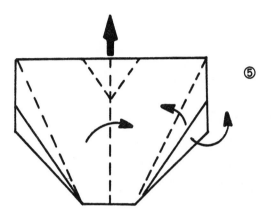

⑤ Turn the plane over and fold up along the dotted lines. Fold back out along the solid lines. Invert-fold up along the lines at the top to make a tentlike lifting flap, which will keep the plane level in flight.

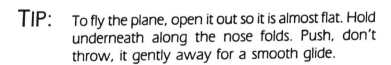

TIP: To fly the plane, open it out so it is almost flat. Hold underneath along the nose folds. Push, don't throw, it gently away for a smooth glide.

☆ SABER-TOOTHED BAT ☆

This plane has its roots with a box-plane design, although its weight distribution has been changed for better stunt flying. With its long fangs out front, this one is sure to turn heads. Try folding the wings out at a slight angle, instead of as shown, to vary the performance of this crazy craft.

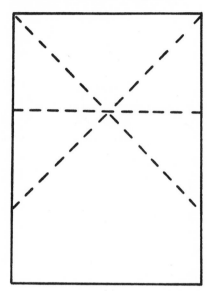

① Begin folding as with other box planes. Fold along the lines, open, and push in the sides as you flatten.

② Fold the bottom corners to the top and crease. Unfold, then bring the sides in to the middle. Make the final crease to the bottom corners as shown.

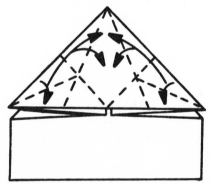

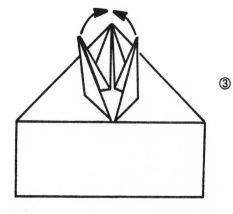

③ Bring the top section together (the sides in to the middle) and fold the bottom section up as shown.

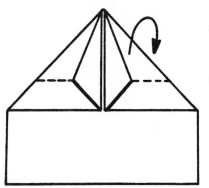

④ Fold the back section of the top away from you as low as possible, as shown by the dotted line.

⑤ You should have two points up front. Fold the plane in half toward you.

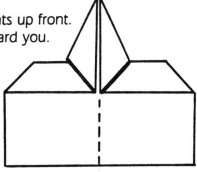

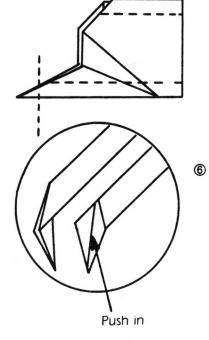

Push in

⑥ Fold out the wings as shown and fold up the stabilizers if desired. You can make the fangs by folding the points in as shown by the exploded view.

TIPS: This plane performs stunts well. You can add tabs to the wing tips for added lift. To gain better balance, fold smaller wings and leave a larger belly.

☆ THE GULL ☆

This plane's name comes from its resemblance to the seagulls as they race through a cool ocean breeze. This is perhaps the easiest birdlike plane to build and it even flies well! A little experimentation with the wing angle will enable it to fly above the waves. Use some of these folds together with your own imagination to invent your own flock.

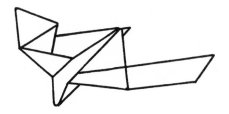

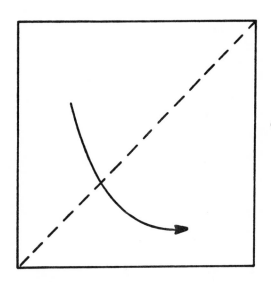

① Start with a square sheet; 8½- x -8½-inches is fine. Fold from corner to corner and crease.

② Fold the folded side over as shown. Make the fold about 1½ inches from the edge.

③ Fold the whole thing in half toward you.

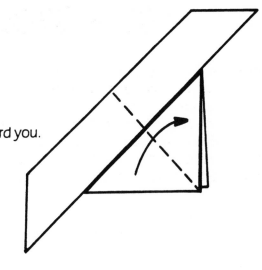

④ Fold the wings back along the line. Make sure it looks like figure 5 when you are done.

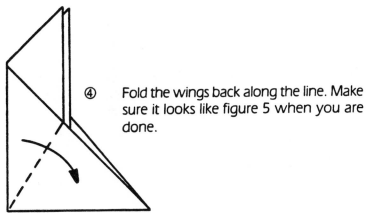

⑤ Fold the wings out along the line. Leave the head a little larger than the tail.

TIPS: The Gull needs to be thrown with some force because of the small wings. Experiment with the wing tips to get a stable flight. Stunts are not out of the question, but they will be difficult to get with this bird.

☆ THE STUNT LOOPER ☆

This one is a somewhat more advanced version of The Looper. It has even better stability. It is perfect for flying outside and can handle almost any kind of breeze. Loops, curves, and even barrel rolls are at your fingertips with this model. A variation, shown in figure 3, will allow you to move the center of gravity a little. Tuck the flap in for flying in the breeze. Leave it out for indoor flight.

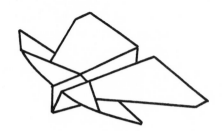

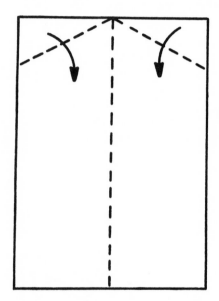

① Fold an 8½-x-11-inch sheet in half lengthwise and open it again. Fold the top corners down as indicated. It is not critical to be exact with this fold, but try to get it as close as possible.

② Now fold along the dotted lines by folding side A to side B and repeating on the other side. Open up. Then fold toward you at the cross of the X (accordian fold).

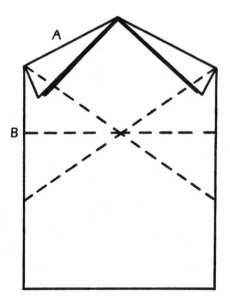

③ Fold the bottom layer of each side in as far as it will go. Most of this fold will be hidden under the top layer. Now if you are throwing this plane outdoors, you might want to fold the flap (A) under and in along the dotted line. This fold adds weight to the nose.

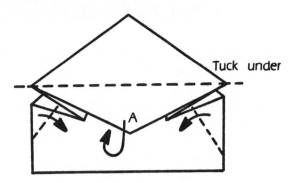

Tuck under

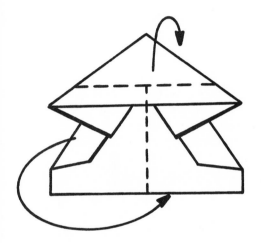

④ Fold the top away from you on the line. Make the width of what is left at least 1 inch. Fold the plane in half away from you.

⑤ Fold down the wings at an angle as shown. The more the angle, the faster it will fly.

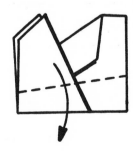

TIPS: Curve up the front and/or back of the plane to give extra lift for loops. Bend one wing tip up for rolls. You can throw this plane hard for a long glide to the ground.

☆ LOCK-BACK JACKKNIFE ☆

We have here another hot performer. It can do 0 to 60 ips*
in a flick of the wrist. It turns on a dime and gets great gas
mileage. You don't even need to take out a loan to own one.
Just follow these directions.

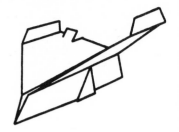

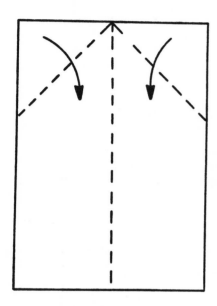

①   Fold an 8½-x-11-inch sheet in half
lengthwise and open back up. Fold the
top corners down to the center.

②   At about ¾ inch below the bottom of
these folds, fold the top section over and
down.

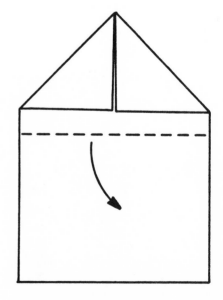

*inches per second

③ Now fold the top corners in to the center again. A little tab should stick out. Fold it up to lock in the other folds. The plane should look like figure 4.

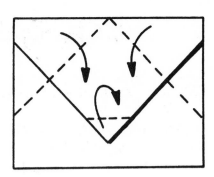

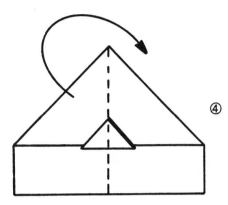

④ Fold the plane in half away from you.

⑤ Now fold the wings out at an angle, as shown.

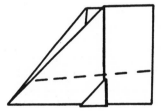

TIP: Bend the wing tips up for extra lift. This plane glides well, and with enough lift, also does loops.

☆ THE FLY ☆

Here is another little model that has good performance and is easy to make. My 9-year-old brother showed me this one. He says that it is very popular on the playground at his school. Try it yourself and see that the younger generation knows their paper airplanes too.

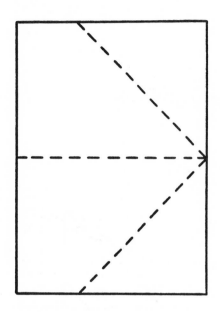

①   Start with an 8½-x-11-inch sheet. Fold it in half widthwise and open again. Fold the corners in to the center.

②   Now fold the point over, to the bottom of the previous folds.

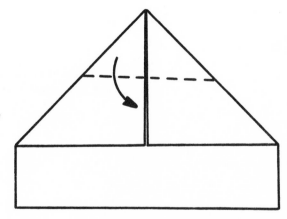

③ Fold the sides in to the center along the lines. Sides A and B should meet at the center.

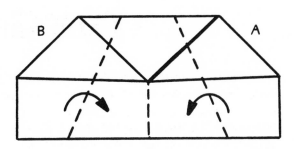

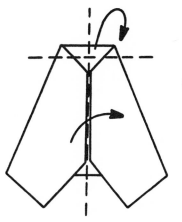

④ Fold the top ½ inch away from you at the top. Crease well; then fold the plane in half toward you.

⑤ Open the wings out as shown.

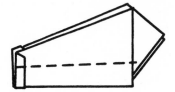

TIPS: For level flight, bend up the wing tips a bit. Bend them up at a 45-degree angle for loops. This model glides best with a firm throw.

☆ THE GREMLIN ☆

The Gremlin takes a turn from the usual folds that make paper planes. The beginning folds can be the start of several variations that all fly well. The weight in the nose (step 3) is important for good balance on all paper aircraft. This is a fun plane to make and to fly. Even if your office is too small for flying, it makes a good addition to your OUT basket.

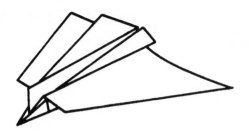

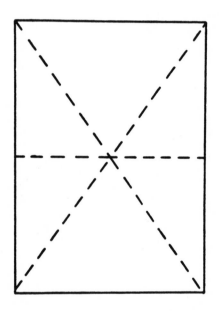

① Start with an 8½- x -11-inch sheet. Make folds from opposite corners as shown. Open it up and fold the sheet in half along its width.

② Now accordian-fold the whole thing, as for the box planes. The two inner flaps will overlap as shown. Fold the top surfaces into the center.

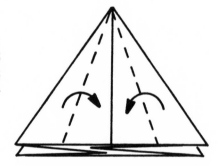

③ About 2 inches down from the point, fold away from you (A). Fold back up at B. Now fold the plane in half to get figure 4.

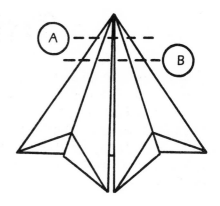

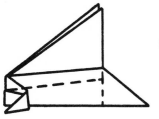

④ Fold the wings back at a slight angle and fold up the wing tips a bit. Your Gremlin should now be ready for all sorts of trouble.

TIP: Wing angle and lift are important for a good flight. With a little practice, you can coax nice loops and rolls out of this plane.

☆ THE HUNTER ☆

Like an arrow, it flies swift and sure to its prey. The Hunter is an unusual and almost old-fashioned approach to getting the weight in the right places. Rolling folds in the nose really give it a straight flight. Use this one for distance events. You can throw it as hard as you like. Try it with heavier paper for even greater distance.

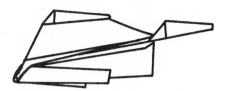

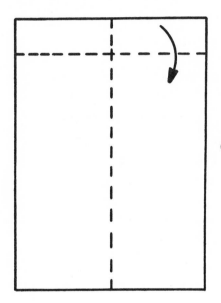

① Fold an 8½-x-11-inch sheet in half lengthwise and open it again. Fold about 2 inches of the top over and crease.

② Fold this section in half and crease. Fold it in half again and push flat.

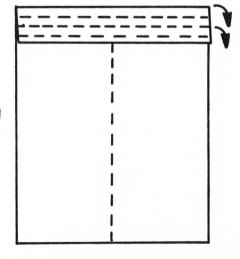

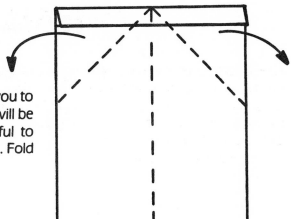

③ Now fold the corners away from you to the center of the back. The nose will be thick, and hard to fold. Be careful to crease this part as well as you can. Fold the plane in half toward you.

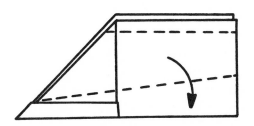

④ Fold out the wings at a slight angle. Lastly, fold up the stabilizers at the edges.

TIP: The Hunter is too long and its nose too heavy for stunts. This is primarily a distance plane. Use a piece of tape to hold the bottom closed and help you obtain a long, smooth flight.

☆ RAM JET ☆

A nice all-around plane, the Ram Jet makes use of some folds seen earlier in this book. I have always found this one well suited for use at large lecture classes, concerts, or ball games. It has always proved to be a crowd pleaser.

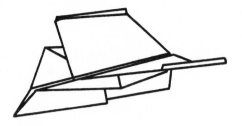

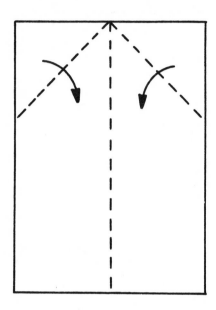

① Use an 8½-x-11-inch sheet. Fold it in half lengthwise and open. Fold the corners in as shown.

② Fold along the dotted lines by bringing side B over to side A and creasing. Repeat on the other side. Accordian-fold the sides in, as for the box planes. Fold the top triangle back up along the line to get figure 3.

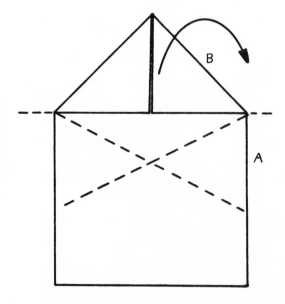

③ Tuck the top surface edges underneath as shown. Turn the plane over.

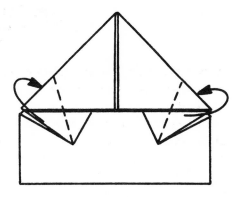

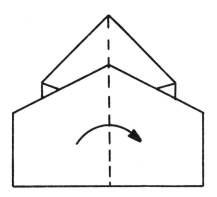

④ Fold the plane in half toward you.

⑤ Fold the wings out about 1 inch from the bottom. No angle is needed here.

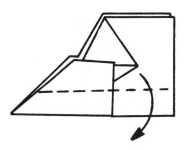

TIPS: You can experiment with stabilizers on the sides of the wings and with tabs on the wing tips. This plane does nice stunts and can also glide long distances.

☆ THE BOAT PLANE ☆

Although it looks almost nothing like a plane, it flies well and is worth your effort. You might not believe it, but it can do loops and circles quite handily. During a level flight, this model plods along slowly because of the large 'bow' up front. It really does look like an old rowboat floating in the air!

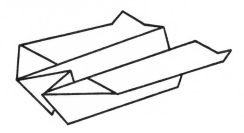

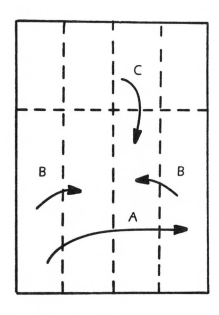

① Fold an 8½- x -11-inch sheet in half (A) and then in quarters (B) by folding each side in to the center. Open it back up and fold the top over about 4 inches (C).

② Fold one corner in a quarter of the way and crease.

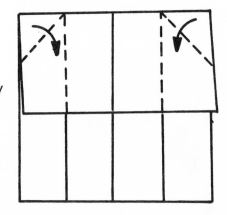

③ Open up completely and push in at A, fold down at B, and fold in the end quarters to the middle. Reverse-fold the top surface. Repeat for the other side. Fold the plane in half.

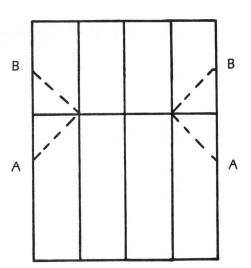

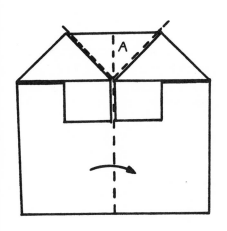

④ Then fold up along the lines, pushing section A up and in. Section A should be inside. Now open the wings out along the fold you have already made.

⑤ Reverse-fold along the nose. Then fold down the wings.

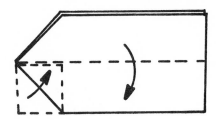

TIPS: The plane should fly open in the middle to give the bow a chance to lift the plane. To do loops, toss the plane up and away.

☆ THE WING ☆

This plane has wings, but no body. These designs have been around quite awhile, and some models are good for long flights. This design, however, uses advanced folding to give it good weight distribution and an improved airfoil. The wing can give you some memorable flights.

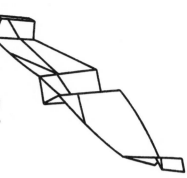

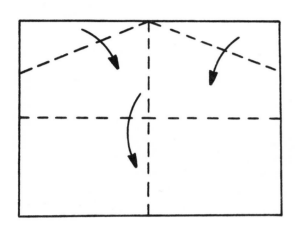

① Use an 8½-x-11-inch sheet. Fold it in half along both its length and width and open back up. Working from the wider side, fold each corner to meet the crease along its width. Now fold the top half to the bottom.

② Fold the top corners to the bottom edge.

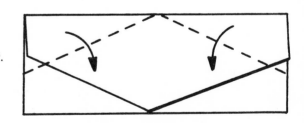

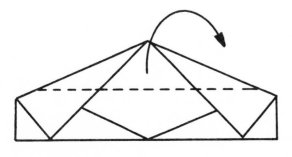

③ Here are two variations. Either fold the top back to the bottom edge . . .

④   . . . or just fold an inch or so back to the other side.

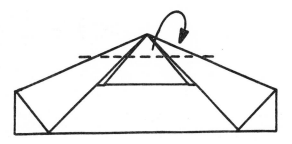

⑤   Now fold the plane in half and fold out the wings on the lines.

⑥   Fold up the stabilizers to complete the wing.

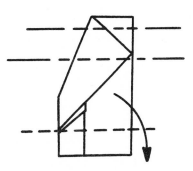

TIPS: Push the wing gently away from you, nearly level. It glides well, but doesn't do stunts.

☆ **SUPER DART** ☆

Here we have a true winner. Super Dart is a great distance flyer. If you construct it properly, you can throw it as hard as you would throw a baseball. I was surprised, as you will be, at flights usually over 100 feet. The front wing gives the plane unparalleled stability.

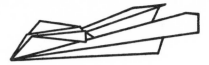

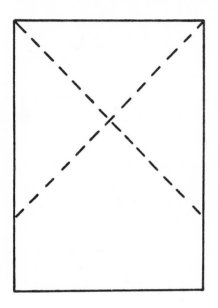

①   Use an 8½-x-11-inch sheet. Fold the corners over as shown, and accordian-fold the sides in.

②   Bring the bottom corners of the triangle to the top.

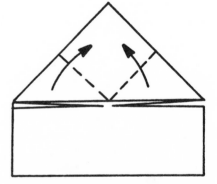

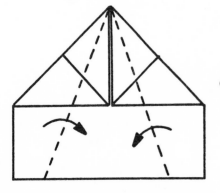

③   Tuck the sides of the plane in to the center, folding as shown.

④ The plane should now look like this. Fold it in half toward you.

⑤ Fold out the wing on the line. Now fold out the front wings so they are on the same angle as the bottom of the plane. This angle gives the needed stability.

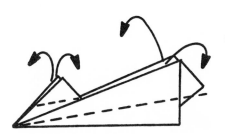

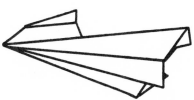

TIP: This is not a stunt plane. It needs a high and hard throw. Distances of 120 feet are not difficult.

☆ BEE BOMBER ☆

The Bee Bomber is so named because of the small legs up at the nose. Once it is in flight, you might notice other similarities as well. It is a good glider and performs stunts better than most box planes. One caution though: it must be made from yellow lined note paper!

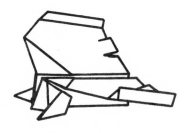

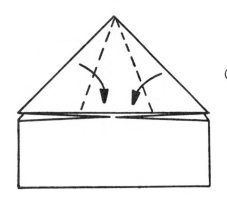

①   Start by making the basic accordian fold (see box planes) on an 8½- x -11-inch sheet. Fold the outer edges in to the center as shown.

②   Fold the top section to the back so that the pointed ends on both sides are even.

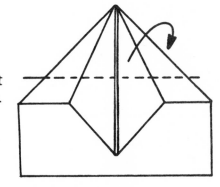

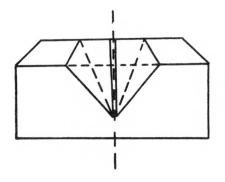

③   Fold along the dotted lines as shown Crease and open back up.

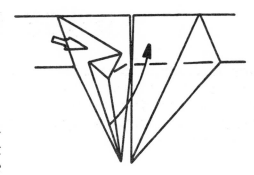

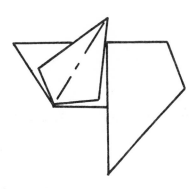

④& ⑤ Push in the top section at the open arrow to the crease you just made, (but don't fold over the bottom part of the section). Fold the point over. Repeat for the other side. The legs should cross a little bit. Now fold the plane in half away from you.

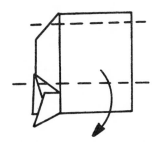

⑥ Fold the wings down even with the bottom. To finish it, fold up the stabilizer flaps at the sides of the wings.

TIPS: Tear tabs up at the wing tips for loops. A medium push will give this Bee the best flight.

☆ X-WING FIGHTER ☆

Here is another awesome-looking craft. Yes, it is Luke Skywalker's favorite. Even though it is not difficult to build, people will certainly ask you how you did it. Upon testing this model, you will find that the X-shaped wing tips change the flight of this fighter a great deal, depending on how they are bent. Good luck.

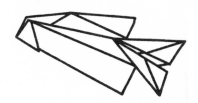

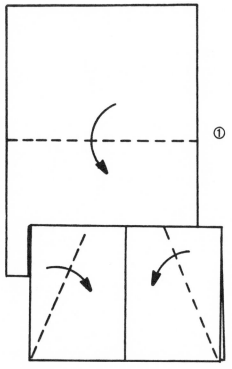

① Fold an 8½-x-11-inch sheet in half widthwise.

② Bring the top corners (of the folded side) in to the center as shown. They will overlap a little. The fold should end in the bottom corners. Accordian-fold the corners inside as with other models.

③ Now fold the new top sides in to the center.

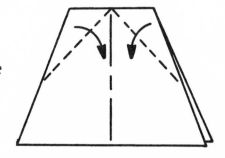

50

④ Fold about 1 inch of the nose back to the other side. Flatten this out well and fold in half toward you.

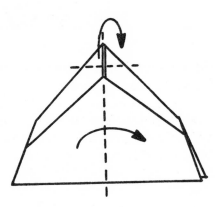

⑤ Fold out the wings at a slight angle and unfold the wing tips. Now you are ready for war.

TIP: This is primarily a glider. Experiment with the wing tips to get a smooth flight.

☆ KING WING ☆

The King Wing is the finest example of a flying airfoil I have yet seen. Its thick leading edges simulate those on real planes very well, and the weight distribution is ideal for a plane with no fuselage. The folds used in this plane hopefully will give you ideas for planes of your own, such as the addition of a tail. This is, by far, the most versatile craft of those without bodies. Notice, also, how the folds lock themselves together at the bottom.

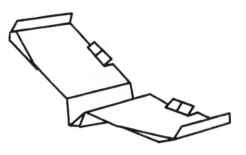

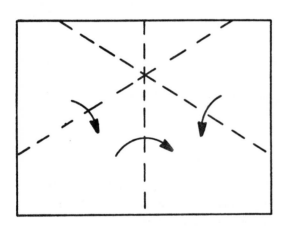

① Fold an 8½-x-11-inch sheet in half widthwise, then open again.

② Bring the corners down to the center of the bottom edge, as shown, and accordian-fold the center section out.

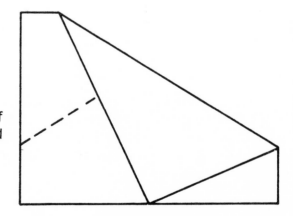

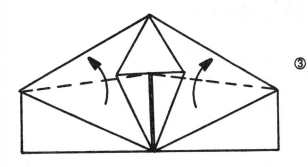

③ It should look like this. Fold the edges at the bottom up along the dotted line as shown.

④ Now fold up the part that sticks out and tuck it underneath. Fold the top section to the back side along line A. Crease well. Flip the plane back over so the folded top side is still in the back.

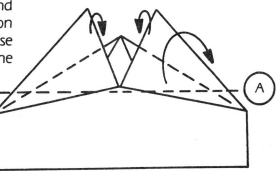

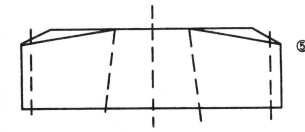

⑤ Fold the plane in half toward you, then fold down the wings and stabilizers along the dotted lines as shown.

TIPS: You should get a fairly level flight without any trim on this plane. Add tabs in the wings to help the lift if needed. You can get small loops by bending the tabs up further.

☆ GLIDER ☆

Here is a fine plane to add to your collection. It is perfect for throwing off balconies at the basketball game or any sporting event. It glides far and long, and is not affected much by winds. The folds used in this design are quite different and can provide some interesting variations of this plane. Note the dual folds in the wings and how they sweep up the wings as they go back. This is another way to provide lift for many planes.

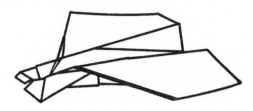

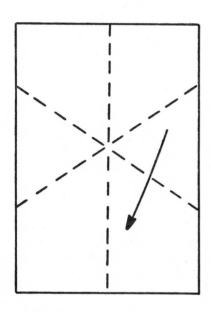

① Start by folding an 8½- x -11-inch sheet in half lengthwise and opening it again.

② Take the top corner halfway between the opposite bottom corner and the middle and fold. Do this for both sides and accordian-fold the sides in.

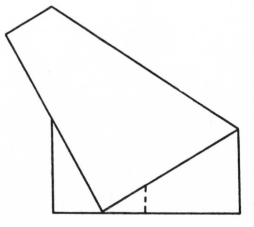

③ Fold the top sides in to the center as shown.

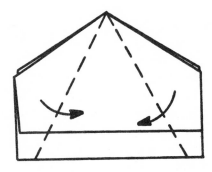

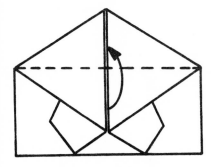

④ Now fold the bottom edges up at the line.

⑤ It should look like this. Fold the points (A) to the center along line B-C by tucking them behind the top surface. Tuck them back again at line D-C and crease. Fold the plane in half away from you. Fold down the nose.

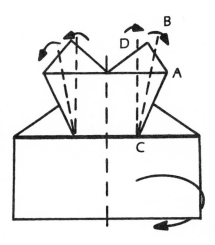

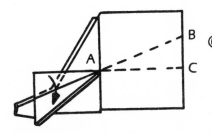

⑥ Fold the wings out along the angled line (A-B) and crease. Now fold down along line A-C. This should make the wings sweep back and up.

TIP: This plane should need no adjustment to fly well. Too bad it doesn't do stunts as well as it glides.

☆ HAWK ☆

The Hawk is a master of high-altitude gliding, as is the real bird. Its wide wings and sharp beak have been folded to give it excellent flying characteristics. Give this bird plenty of room and let it soar.

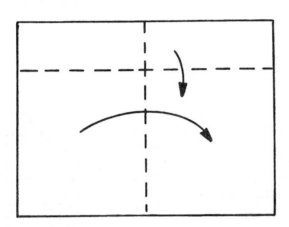

① Begin with an 8½- x -11-inch sheet and fold the long edge over about 2 inches as shown. Now fold it in half widthwise.

② Fold the topside back to the edge as shown. Repeat on the other side. Open the paper back out and make accordian-folds to get figure 3.

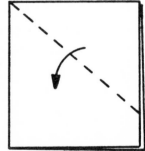

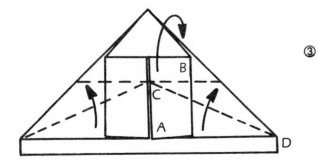

③ Fold up corners A along line C-D so they are even with the diagonal edges. Tuck the corners under the flap of the top triangle at B. Now fold the top to the back side along the line as shown.

④ Turn the plane over; it should look like this. Crease along the dotted lines and bring the bottom point up as shown. The creased folds should invert-fold as the point is folded up. This is tricky so go slowly.

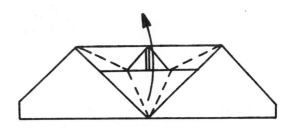

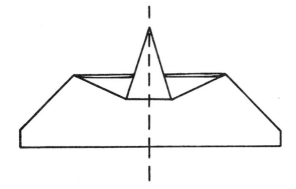

⑤ Fold the plane in half away from you.

⑥ Fold out the wings and fold down the stabilizers where shown. Now invert-fold the beak down and, with your finger inside, crease down.

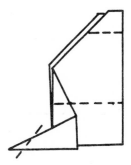

TIPS: Add tabs at the back of the wings and experiment with them for a glide or a loop. Good luck.

☆SUPER-MODIFIED BOX PLANE☆

This is an example of what you can come up with by using your imagination. Here is a plane that can accomplish any stunt with ease and is also a long-range glider. The tail will allow you to make changes in lift and trim in a moment. All you need do is change the angle of the small flaps at the end. You will also see the added stability that a tail will give to this design and many others. Have fun with this one!

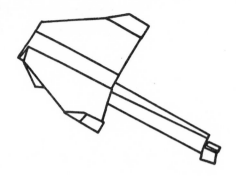

① Start with an 8½-x-11-inch sheet and fold over the top 4 inches (just guess at this distance).

② Now fold the top sides in to the center. Now fold up about 1½ inches from the bottom, as shown, and crease well. Tear this strip off to use as the tail.

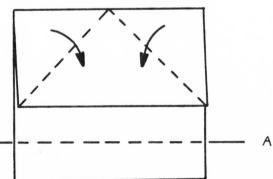

③ Fold the top corners down in to get an accordian fold as shown. Now fold the bottom corners of the top piece in to the center along the lines.

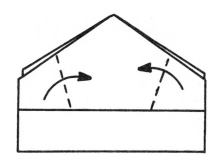

④ Fold the top down over the previous folds at line A, then open to insert the tail section. Push the tail all the way up to the nose. Now refold the nose over to lock in the tail. Fold the top point back out toward the front of the plane. This adds a little more weight to the nose. Tear the tail in a bit at B for a control surface, then fold the whole thing in half away from you.

⑤ It should look like this now. Fold the wings out and add stabilizers if you wish. Fold out the little wings at the end of the tail.

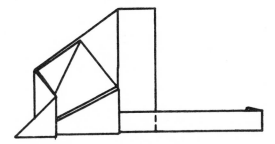

TIP: Use the little wings on the tail for lift and directional control. They work well, and you don't need to bend them far to change the flight.

☆ SST ☆

The SST is a cousin of the Undercarriage Plane and flies just as well. It uses a very unusual locking fold in the nose, which also gives it good balance. This is a long-range jet. Maybe it can make it across your big pond.

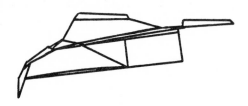

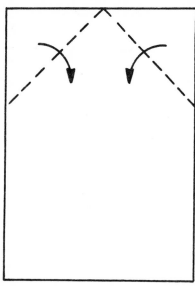

① Use an 8½- x -11-inch sheet. Fold the top sides to the center.

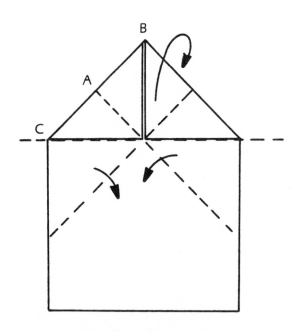

② Begin by folding side A in half, by folding the top corner (B) to its bottom corner (C). Keep its edge even. Repeat on the other side. Open these folds back out (they will be used later) and fold the triangle away from you to the back side. Turn the plane over.

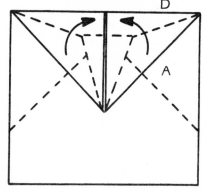

③ Carefully fold up along the dotted line where shown by bending the diagonal side to the center then to the top (side D). Push up at A.

④ Fold the point up and over the pocket, pushing in at the arrows. It should flatten out nicely.

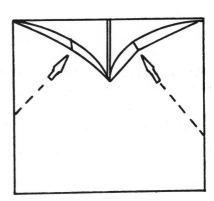

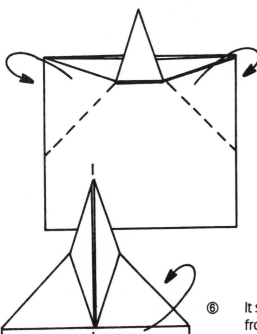

⑤ Now fold away from you on the creases made earlier. This locks the nose in place. Turn the plane over.

⑥ It should look like this. Fold it in half away from you.

⑦ Fold the wings out as indicated. You can push open the nose and make an inverted fold to complete the SST look.

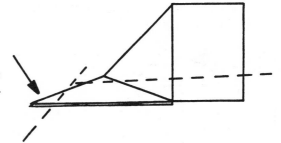

TIPS: Bend up the wing tips a bit for lift. Throw strongly for a long glide.

☆ LONG-DISTANCE CHAMP ☆

Last, but not least, is my favorite distance plane. It is complicated, but worth it. It is super stable and perfectly balanced. Throw it as hard as you can and watch it go out of sight. Distances of 150 feet can be obtained with this true champion. The small wings up front add just enough stability and lift, without adding drag. Bend the front tips up to level out its flight.

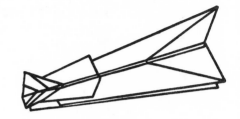

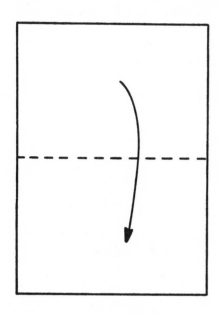

① Start with an 8½-x-11-inch sheet; fold in half widthwise as shown.

② Fold the top corners in to the center. Open and push the sides in to make an accordian fold.

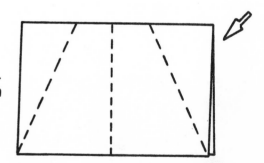

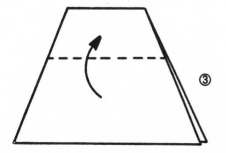

③ Fold the top piece back along the line.

④ This is the tough part. Lay the plane flat, take the corners at A and B, and *slide* them up to the middle. Keep your thumbs on the lower part of the plane. The top center section should bow out as you do this. Hold them at the middle and crease along the short lines (C).

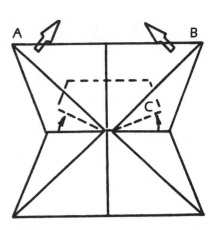

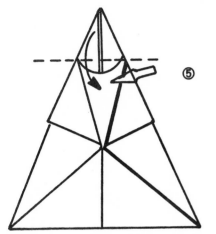

⑤ Push the center part flat and fold the two points over to get figure 6. The points will overlap.

⑥ Fold the plane in half toward you.

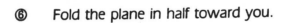

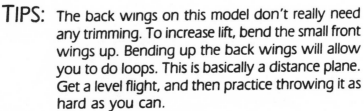

⑦ Fold the wings out on a slight angle as shown.

TIPS: The back wings on this model don't really need any trimming. To increase lift, bend the small front wings up. Bending up the back wings will allow you to do loops. This is basically a distance plane. Get a level flight, and then practice throwing it as hard as you can.

NOVELTY PLANES

This section is devoted to ingenious and interesting aircraft, each with a unique design. High performance is not important here, although all the planes fly. The designs might spur you onto new designs for planes—both paper planes and real aircraft. Try them. They are amusing and will, at least, occupy those lazy times at the office or at school.

☆ TAIL PLANE ☆

This is a plane from my childhood, and perhaps from yours, too. In grade school, I made more of these than any other design, just because it was different. It flies well as a glider and is fun to assemble. Have fun with this oldie but goodie.

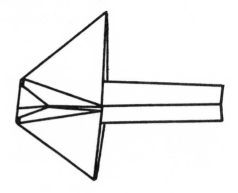

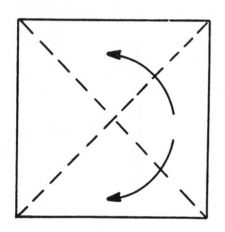

①   Start with an 8½-x-11-inch sheet and make a square sheet from it. Save the strip left over for the tail. Accordian-fold the square sheet as shown.

②   Fold the bottom corners up to the top corner.

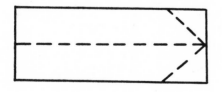

③   Fold the tail section in half lengthwise. Then, fold the top corners of the tail section.

④ Push the tail in the middle of the other piece. Fold the top sides of the top sheet in to the center and fold the top over them.

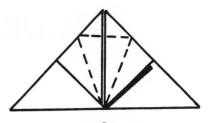

⑤ Open up the slots in the top section and tuck the corners in as shown.

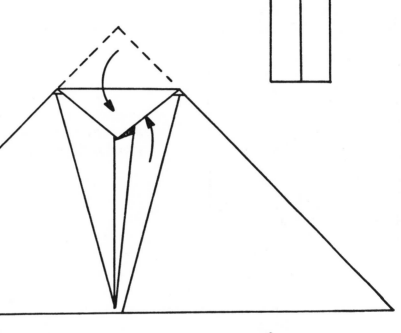

⑥ Fold the plane down the middle, then open back up. Launch with a smooth push.

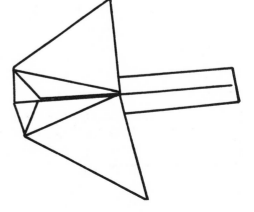

☆ MODIFIED TAIL PLANE ☆

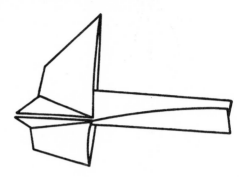

This is a variation of the Tail Plane for those who prefer a more complicated plane of this sort.

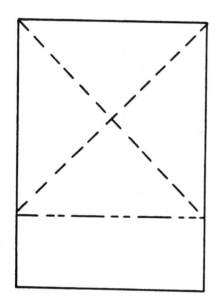

①   Start as with the tail plane, saving the extra strip of paper.

②   Fold the bottom corners up to the top corner.

③   Fold the top edges in to the middle, as shown, and open back out. Follow by folding the bottom edges up to the middle.

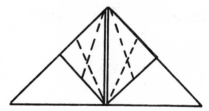

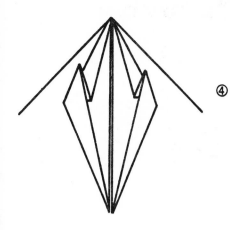

④ Pull out the bottom section while folding the top section back down then push the bottom section back up, folding the top underneath the bottom. It should look like figure 5.

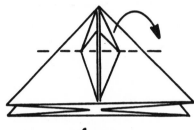

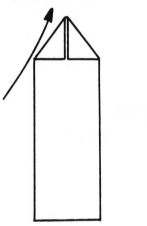

⑤ Insert the folded tail section in the middle of the wings. Fold the top section back to the other side.

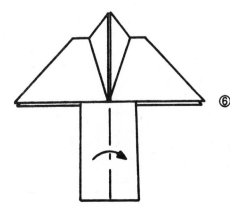

⑥ Fold the plane down the middle a bit, and you are ready to go.

☆ SPIN KING ☆

I want to make sure that this is a complete book of paper aircraft, and I have been careful to remember this silly plane. There are times when graceful flights and smooth landings are upstaged by the crazy antics of this plane. Although it actually flies well, it gets from one place to another spinning madly. Usually at the end of its flight, it levels out as its airspeed decreases. With a good toss, however, it might spin out of sight.

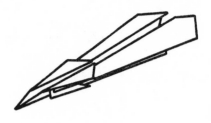

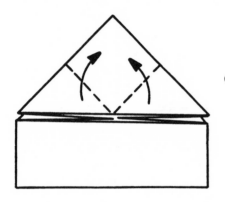

① Start with an 8½-x-11-inch piece and accordian-fold the top section as with the box planes. Fold the bottom corners up to the top corners.

② Fold the top sides of the diamond to the middle and crease. Open back out again to a triangle.

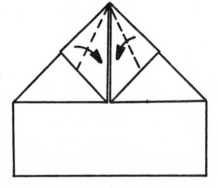

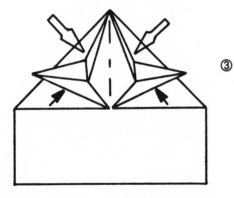

③ Reverse-fold along the bottom creases made previously by pushing the upper sections in first and folding the bottom corners over and up. It should look like figure 4.

④ Fold the bottom surface in and under so that the edges meet at the center as shown. Now fold about 1½ inches of the top away from you to the back side, leaving the two points sticking out.

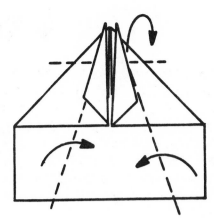

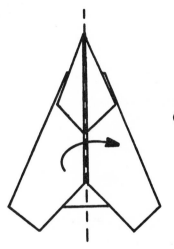

⑤ Fold the plane in half toward you.

⑥ Fold the wings out as shown along the lines. Throw high and hard.

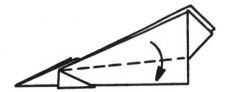

☆ THE CIRCULAR AIRFOIL ☆

A flying tube? If you don't believe it, be sure to try this one first. It really flies! The leading edges actually create the necessary lift. You can get different flying characteristics from tubes of different lengths and diameters. Try a few variations to see which one you like best. This model should definitely win the Originality award at your hometown contests.

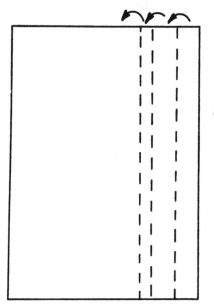

① Use an 8½-×-11-inch sheet or vary the size to suit you. Long, thin models do well for distance. Fold over about 1 inch on the longer side. Fold over again. Now fold this in half.

② With the folds to the outside, pull the sheet across the edge of a table several times to curl the paper.

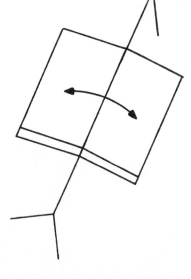

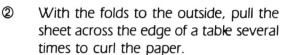

③ With several pieces of tape, construct a tube as shown. Hold the plane like a large glass and throw it gently, letting it roll off your fingertips.

☆ FLAT FLYER ☆

Here is a pure and simple airfoil with almost nothing to it. If you build one very small with light tissue paper you might get flights over 10 seconds. Launch it from over your head with a light, forward push.

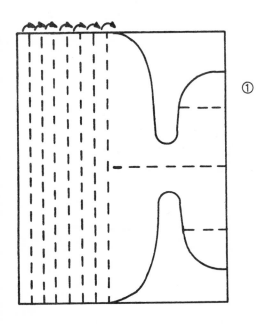

① To make a competition-winning stunt plane, start folding on the long edge, as shown. Fold in about 3½ inches, with ½-inch folds as shown. With a pair of scissors, cut out a shape like the one shown. Fold the wings out and the tails down. No kidding, this may prove to be the best flying plane in the book. It will do stunts better than any model I know of. Try it!

☆ MAJOR ARROW ☆

A long-distance glider is all the claim I make on this model. It flies straight as an arrow. Although it does not create a great deal of lift, necessary for gliding, it can be thrown 100 feet or more. It can be a lot of fun throwing across a crowded room. One trick with this plane is quite effective: cut a notch in the nose of the plane (see figure 7) and use a rubber band to catapult this arrow into the sky!

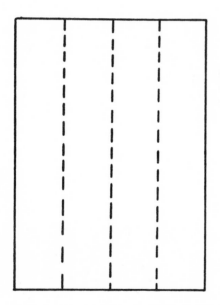

① Fold an 8½- x -11-inch sheet in half lengthwise then open. Bring the sides to the middle and crease as shown.

② Now fold the corners away from you to the back. Follow by turning the plane over to get figure 3.

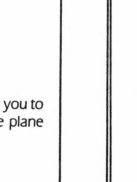

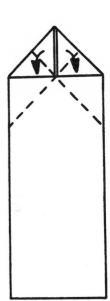

③ Fold the upper triangle sides in half and crease along the lines. Open up to the rectangle (figure 2) and accordian-fold the bottom sides. Next open up the sides of the triangle, fold the bottom edges up to the middle and fold the top triangles down over them, flattening the small diamond shape (again like an accordian fold).

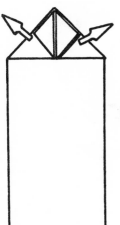

④ With your finger at the top of the diamond bring the bottom flaps out until it looks like figure 5. Flatten this out well.

⑤ Now fold the sides back to the other side along the lines. This will lock the nose together. Turn the plane over.

⑥   Fold it in half toward you.

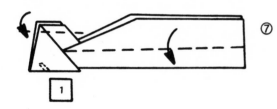

⑦   Fold out the wings along the dotted line
and fold out the canard wings to finish
the arrow. Cut the notch at **1** to use
a rubber band to power the flight.

☆ HELICOPTER ☆

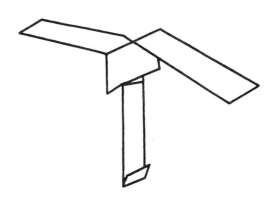

The paper Helicopter is an age-old toy that both the young and old can enjoy today. Its flight is quite unique, and it can cover a great deal of ground if launched from a good height. Try throwing one off your roof or out of a window.

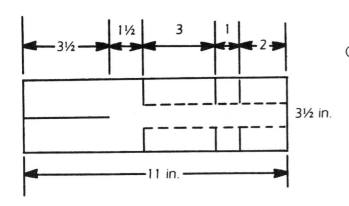

① Tear a 3-inch-wide strip from an 8½- × -11-inch sheet and cut as shown by the solid lines. Fold the 3-inch and the 2-inch sections in to the center, overlapping to get figure 2.

② Starting at A, fold the bottom over on the line; then fold again. Tuck the tabs into the slots you have made. Finally fold down the wings—one to the front, the other to the back—to get the finished model. Drop from a high place and watch it hover to the ground.

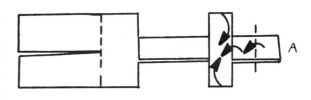

☆ EAGLE ☆

The discipline of origami usually attempts to pattern its creations after nature. Paper trees, caterpillars, and butterflies are standard origami creations. Paper birds are also very popular. Because of the strange shapes of paper and intricate scissor work frequently used, many of the designs have been excluded from this guide. Several of these complex creations fly quite well. The Eagle is a simple example of a flying origami design. With a little experimenting on the trim of this plane, you too can fly like an eagle.

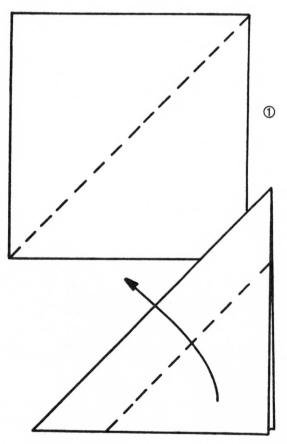

① Start with an 8½- x -8½-inch sheet (made from an 8½- x -11-inch piece). Fold it in half, corner to corner.

② Fold back both sheets as shown. Leave about 2 inches for the wings.

③ Now fold back one sheet. Pick the line so that when it is folded you are left with a square showing as in figure 4.

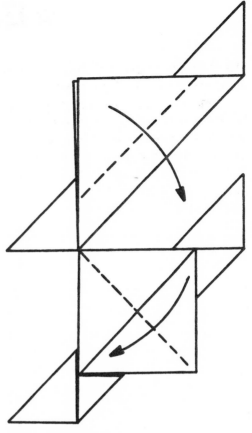

④ Now fold the plane in half.

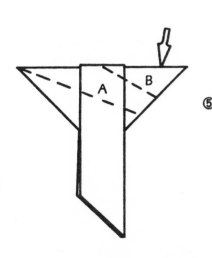

⑤ Fold up on the wings along dotted line A. Crease along line B and push the tail down with an inverted fold. Open this back out.

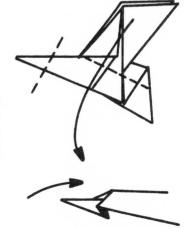

⑥&⑦ It should look like this. Now fold out the wings as shown. Carefully invert-fold the nose as shown to make the beak. With a little bending of the wings, it should fly nicely.

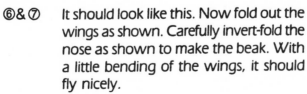

☆ DOVE ☆

Here is another origami bird. It even flies fairly well. Besides, you can always make one for your sweetheart if you need to make up after an argument.

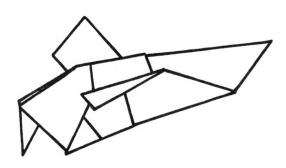

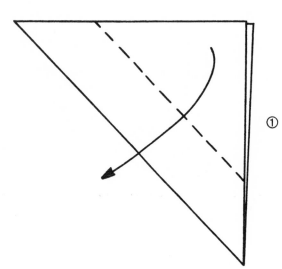

① Start as with the Eagle. Fold the edges back as shown.

② Fold the top sheet back, leaving a square on top.

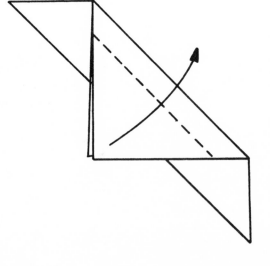

③   Fold the dove in half.

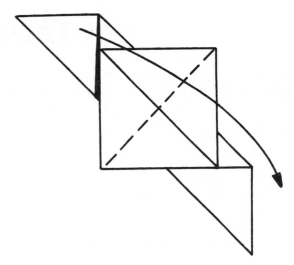

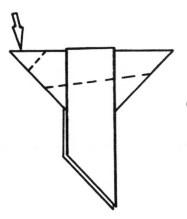

④   Fold the wings up along the line. Push the nose in and down using an inverted fold.

⑤   Now fold the wings out along the line shown. Your Dove should be ready to carry a message of peace.

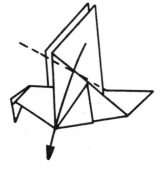

☆ DRAGONFLY ☆

The dragonfly is one of the most amazing flying creatures ever created. To watch one hover and dart among the cattails is sheer poetry. Scientists today are still not in complete agreement as to the mechanics of its flight. The replica shown here is, by far, the most complicated model in this book. Coincidentally, it is also a fine example of advanced origami. Take your time, study the drawings, and you should have a model that is well worth your effort.

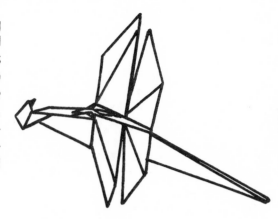

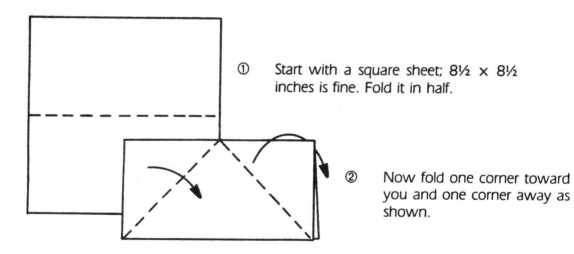

① Start with a square sheet; 8½ × 8½ inches is fine. Fold it in half.

② Now fold one corner toward you and one corner away as shown.

③ Now grasp the triangle at the center and open it up to form the diamond in figure 4.

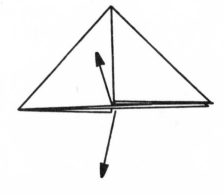

④ Fold along the lines as shown. Carefully
& ⑤ pull the top sheet up and over until the
sides fold in. Crease down the sides. Repeat on the other side. This is a standard origami base fold. It will be used
again for the Bird.

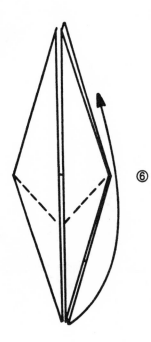

⑥ Study the drawing, then bring the back
surface up to the back side.

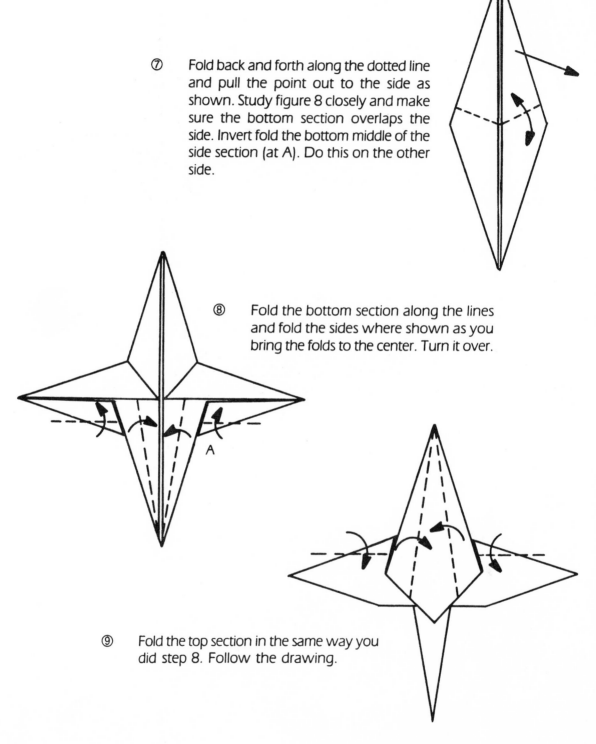

⑦ Fold back and forth along the dotted line and pull the point out to the side as shown. Study figure 8 closely and make sure the bottom section overlaps the side. Invert fold the bottom middle of the side section (at A). Do this on the other side.

⑧ Fold the bottom section along the lines and fold the sides where shown as you bring the folds to the center. Turn it over.

A

⑨ Fold the top section in the same way you did step 8. Follow the drawing.

10 Fold your Dragonfly in half toward you.

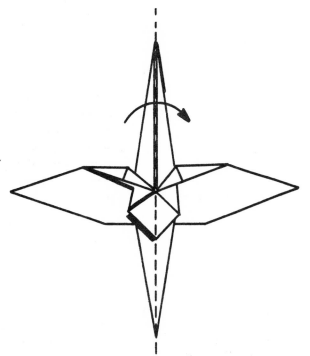

11 Reverse-fold the neck as shown, and fold the head back up and in. Cut the wings if you like to finish the Dragonfly.

TOYS,
TOYS, TOYS

This section has been included for the benefit of anyone who enjoys little gadgets and toys in their spare time. These will not fly, but are interesting in their own way. Many of them you will recall from your own childhood. Happy folding!

PART 3

☆ THE BOX ☆

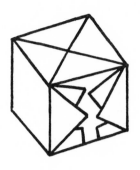

This is an unusual little box that was rumored to contain magic after it was blown up. The rumor may or may not be true, but don't tell your children. They usually love this little box.

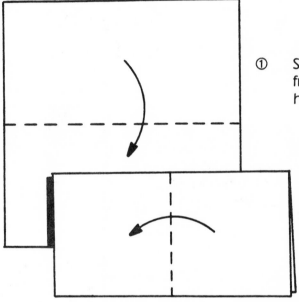

① Start with an 8½- x -8½-inch sheet (made from an 8½- x -11-inch piece). Fold it in half.

② Now fold it in half again.

③ Fold it diagonally. Open it up to look like figure 2, and make an accordian fold, as for a box plane.

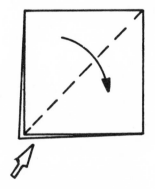

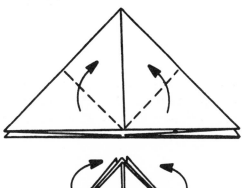

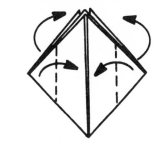

④
&⑤   Now fold the bottom corners up to the top. Turn it over and repeat the procedure.

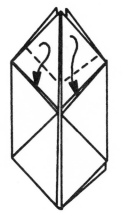

⑥   Fold the sides in to the center. Then tuck the flaps at the top down into the slots you see in the sides. Do this on all four sides. Flatten the whole box down.

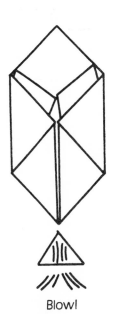

⑦   To open up the box, find the hole at the end and blow hard, inflating the box. To make balloons, try using a whole sheet of newspaper. They are great for parties.

Blow!

☆ **FROG** ☆

This toy is sure to get a laugh from your friends, and at least a chuckle from your boss. It hops across your desk quite nicely, and would be even more effective if your memo pads were made of green paper. Young children will love this little ditty.

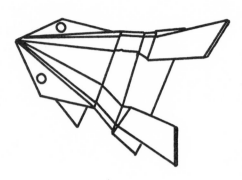

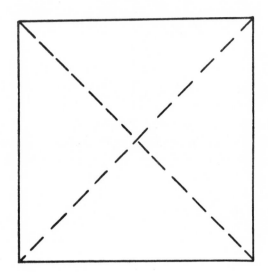

①   Fold a square sheet as shown. Use 8½-x-8½-inch or smaller sheets. Bring the sides in to make a standard accordian fold.

②   Bring the bottom corners up to the top, crease. Turn the Frog over.

③   Fold the sides in to the center.

④ Fold the front feet away from you (A). Now fold the inner edges away from the center on the dotted line (B). This makes the back feet.

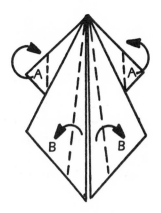

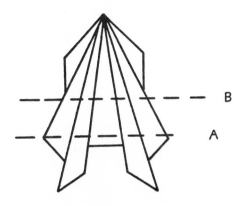

⑤ Fold up at A and crease, then fold in the opposite direction at B. To make the Frog jump, push down along B then slip your finger back. The Frog should jump forward.

☆ PAC-MAN MOUTH ☆

Whether you draw a happy face or your boss' portrait on this toy, it will certainly provide some fun. Just having one of these on your desk will let everyone know that you still have a little youthful spark left. Make them for place cards at your child's next party.

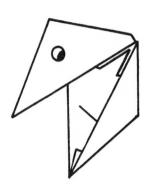

①   Use an 8½- x -11-inch sheet for this one. Begin by folding in half widthwise.

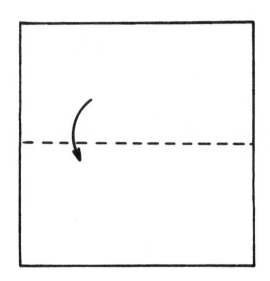

②   Fold the top corners down to the center. Now fold one bottom edge up and over one of the flaps. Fold the other edge up to the back.

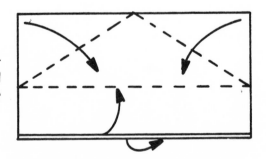

③   Fold the bottom triangles around the top triangles, the front ones to the back and the back ones to the front.

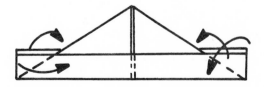

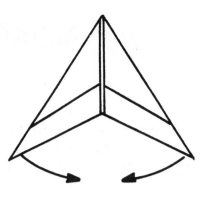

④ Turn the triangle to its side and flip to hold the triangle with the open edge facing you. Placing your thumbs at the center in the opening, fold the top to the bottom, pulling out at the sides to open them out. Flatten the diamond down.

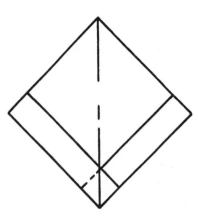

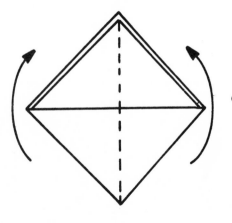

⑤ Fold up the bottom corner to the top at front and back.

⑥ Repeat step 4.

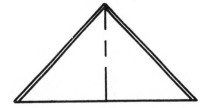

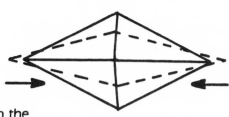

⑦ Open the sides out and fold down the middle piece to either side.

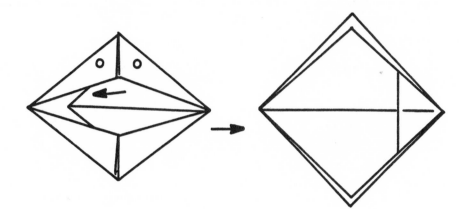

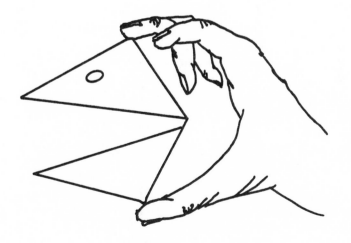

⑧ Bend one side to meet the other and flip. Hold the mouth at the top and bottom corners. Open and close.

☆ HATS ☆

Here are two different hats that can be used for any occasion. The pictures should be self-explanatory. You can use sheets of newspaper to make them in an extra large size.

☆ THE FORTUNE TELLER ☆

This is a very old toy used by school children everywhere. The game is played by picking numbers on the square as it is opened and closed. Then the inside flap is opened to reveal the fortune.

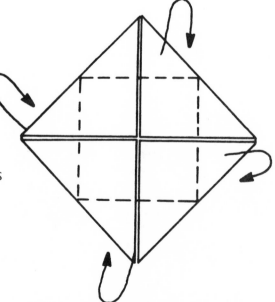

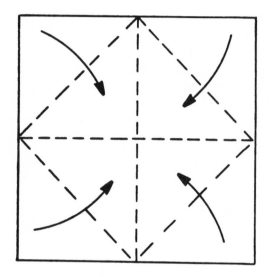

① Start with a square sheet; 8½ × 8½ inches is fine. Fold in quarters and then open up. Now fold the corners in to the center.

② Turn over. Once again, fold the corners to the center.

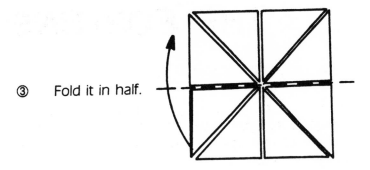

③   Fold it in half.

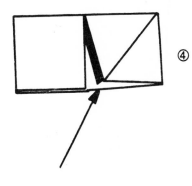

④   Stick four fingers in the flaps at the bottom and push to the center. Open it up to make your finished Fortune Teller. Write numbers on each triangle and the fortunes under them.

☆FOX FACE☆

Here is a cute finger puppet for young and old alike. Perfect for playing the wolf!

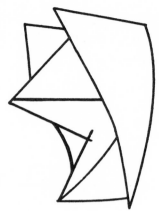

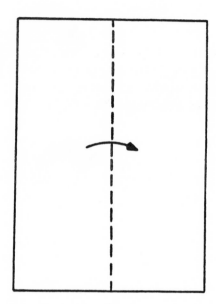

①   Start with a sheet 8½ x 11 inches or larger. Fold in half lengthwise.

②   Open back out and fold in half width-wise. Then fold the top and bottom sides to the middle as shown.

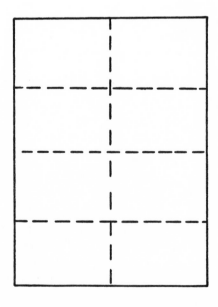

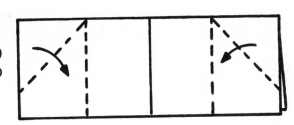

③ Open back out and turn the paper 90 degrees. Refold lengthwise. Fold the top edges in to meet the quarter folds.

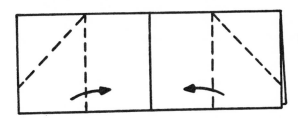

④ Open the corners back out. Then fold the side edges of the top piece to the middle as shown. Fold the triangles back down on top. It should look like figure 5.

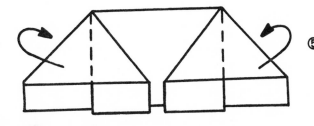

⑤ Fold the edge quarters to the back as shown. It should look like figure 6.

⑥ Fold the bottom of the top piece to the front and the bottom of the back piece to the back as shown. Open slightly and push in at the top. Now all you need are three little pigs.

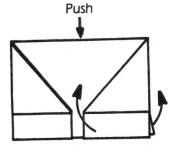

☆ **BIRD** ☆

The Bird is a classic origami creation and has probably been handed down from generation to generation for many hundreds of years. Although it does not fly, it will flap its wings on command. Experiment with this design. You might come up with a crane or a heron, depending on how you fold the neck and tail sections.

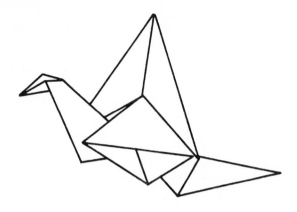

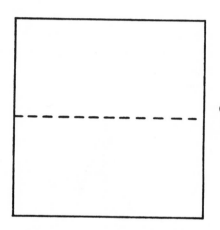

①   Use an 8½-×-8½-inch sheet. Fold it in half.

②   Fold one corner to the center toward you and one to the back center, as you did for the Dragonfly.

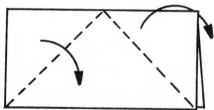

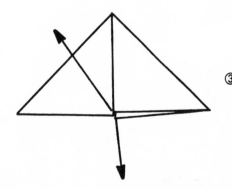

③   Pull this apart to get the diamond in figure 4.

④ Fold along the lines.

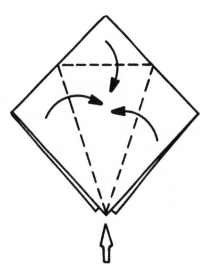

⑤ Pull the top sheet up as shown. Flatten it out. Repeat on the back side.

⑥ Fold along the lines and invert-fold the tail and neck out to the sides.

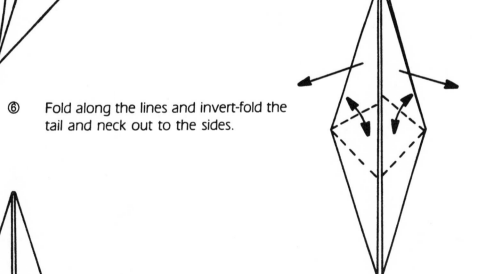

⑦ Fold out the wings along the dotted line and invert-fold the beak as shown at the open arrow. By pulling at the circles the bird will flap its wings.

☆ STAR ☆

The Star is a neat little paper folding exercise that can teach you more about locking mechanisms. Children enjoy making them for decorations and Christmas ornaments. You might have an employee who deserves a star for finishing up that contract before the deadline. Enjoy.

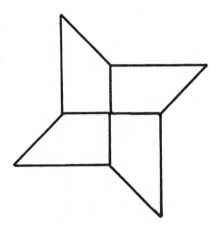

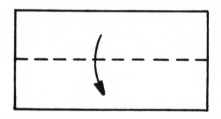

① You will need two pieces of paper for this one. Get them by tearing an 8½-inch-square sheet in half.

② Fold each sheet in half lengthwise.

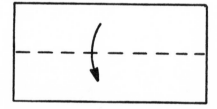

③ Fold them in half. Then fold the top corners to the bottom edge and the bottom edge up to the center, as shown by the drawing.

④ Study the figures closely and arrange the two pieces exactly as shown. Place piece A on top of piece B as shown. Fold the tabs over and tuck them into the appropriate slots. Try making the Star with different-sized pieces of paper.

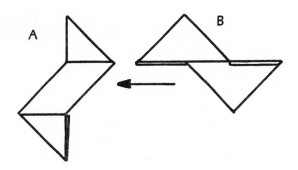

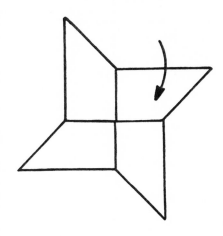

☆ PINWHEEL ☆

Lastly I give you the directions for making your own Pinwheel. No matter what your age or profession, a Pinwheel is guaranteed to bring out the child in you. Spinning in the breeze, it can raise the spirits of all, reminding us that the carefree wonders of youth have only been covered lightly by the dusts of time. I truly hope this book has brought you a few minutes of playful diversion and perhaps a smile or two; for that, after all, was my only intention.

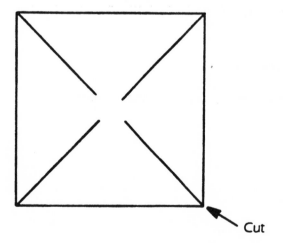

Cut

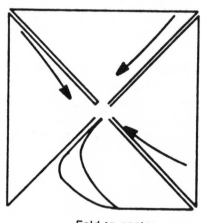

Fold to center
hold with pin

Other Bestsellers of Related Interest

Model Railroad Scenery and Detailing
—Albert A. Sorensen
Build economical, professional-looking layouts following the time- and money-saving advice in this handbook. It contains all the clear, step-by-step instructions and in-progress illustrations you need to plan and detail a realistic, fully operational model railroad in a space even smaller homes and apartments can spare. Checklists ensure you leave nothing out.
0-07-156713-5 $19.95 Paper

Oddballs, Wing-Flappers & Spinners: Great Paper Airplanes
—John R. Bringhurst
A step-by-step guide to the quixotic pursuit of crafting the perfect paper airplane—from an expert! No materials other than paper required. Illustrated.
0-07-067910-X $9.95 Paper

Planes, Jets & Helicopters: Great Paper Airplanes
—John Bringhurst
Fold-by-fold, illustrated instructions for 25 creative designs. Makes building paper airplanes easy . . . even for children.
0-07-007904-8 $9.95 Paper

How to Order

 Call 1-800-822-8158
24 hours a day,
7 days a week
in U.S. and Canada

 Mail this coupon to:
McGraw-Hill, Inc.
P.O. Box 182067
Columbus, OH 43218-2607

 Fax your order to:
614-759-3644

 EMAIL
70007.1531@COMPUSERVE.COM
COMPUSERVE: GO MH

Shipping and Handling Charges

Order Amount	Within U.S.	Outside U.S.
Less than $15	$3.50	$5.50
$15.00 - $24.99	$4.00	$6.00
$25.00 - $49.99	$5.00	$7.00
$50.00 - $74.49	$6.00	$8.00
$75.00 - and up	$7.00	$9.00

EASY ORDER FORM—
SATISFACTION GUARANTEED

Ship to:

Name _____

Address _____

City/State/Zip _____

Daytime Telephone No. _____

Thank you for your order!

ITEM NO.	QUANTITY	AMT.

Method of Payment:

☐ Check or money order enclosed (payable to McGraw-Hill)

☐ DISCOVER ☐ AMERICAN EXPRESS Cards

☐ VISA ☐ MasterCard

Shipping & Handling charge from chart below	
Subtotal	
Please add applicable state & local sales tax	
TOTAL	

Account No. [ ][ ][ ][ ][ ][ ][ ][ ][ ][ ][ ][ ][ ][ ][ ][ ]

Signature _____ Exp. Date _____
Order invalid without signature

In a hurry? Call 1-800-822-8158 anytime, day or night, or visit your local bookstore.

Key = BC95ZZA